The Strategic Security Risk Report

Jayshree Pandya

Jayshree Pandya

ISBN: 9798706567743

Content

The Strategic Security Risk Report

The Strategic Security Risk Report 2021 is published by Risk Group. The information in this Risk Report, or on which this report is based, has been obtained from publicly available sources, Risk Group's research, scenario visualization, and Risk Roundup discussions.

The Risk Group research and reporting of the strategic security risks information makes no representation or warranty, express or implied, as many variables play a role in the onset of strategic security risks. This strategic security risk report's statements may forecast future security events based on certain scenario visualizations and assumptions made today by Risk Group. These forecasting statements based on Risk Group analysis involve many known and unknown strategic security risks, uncertainties, and other emerging factors from the human ecosystem.

While a sincere effort is made to research and report all security risks with strategic security impact, the risk research initiative is a work in progress. Future revisions will likely be made available yearly as Risk Group research reviews more risks in a continually changing human environment.

As a result, readers are cautioned not to place undue reliance on these strategic security forecasting statements. Risk Group will not be liable for any loss or damage arising from using the strategic security risk review information in this report.

Message From Risk Group

In this universe's vastness, it is critical to understand how the reality of human existence in the discrete individual unit impacts all collective living and nonliving systems in all known and unknown domains and dimensions. Our emerging understanding of the connected human ecosystem reveals that known/unknown domains and dimensions have an intrinsic reliance on each other, despite any three-dimensional or four-dimensional distance between the spaces. It implies that even if the domains and dimensions are physically detached with no traditional or measurable linkages, what happens in one domain has a quantifiable effect on the rest of the domains and dimensions.

Along with humans, all living and nonliving beings are part of an entangled global human ecosystem. The ability to generate and manipulate constructive and destructive in/action is at the foundation of disorder across domains. The human ecosystem has been defined by five domains: quantumspace, cyberspace, aquaspace, geospace, and space, based on Risk Group's framework. According to the superstring theory framework, the universe's human dimensions are understood to be in ten different dimensions. Therefore, it is crucial to understand and evaluate how quantum mechanics principles translate to humanity's survival and security. If an individual human in/action is seen as a single atom, is our behavior guided by deterministic laws that impact the collective human species?

Since our individual behavior is a way to gauge what we are doing in this world and to this world, this is a time to brush up on our individual visions and make them a life force for our collective security beyond the human QCAGS (quantumspace, cyberspace, aquaspace, geospace, and space) ecosystem. Let us commit to act in a way that creates a wave of connected security particles beyond human existence.

About Risk Group

Risk Group: Risk Group is a Strategic Security Risk Research Organization, Platform, and Community. Risk Group's Strategic Security Community and Ecosystem is an emerging cross-disciplinary and collective community. It comprises top scientists, security professionals, thought leaders, entrepreneurs, philanthropists, policymakers, and academic institutions from across nations collaborating to research, review, rate, and report strategic security risks to protect humanity's future.

Risk Roundup: Risk Roundup, a global initiative launched by Risk Group, has finished four successful years and 250 episodes of security risk reporting of risks emerging from existing and emerging technologies, technology convergence, and transformation happening across cyberspace, aquaspace, geospace, and space (CAGS). As we enter the fifth year of our collective risk research, review, rating, and reporting of security risks, we are mindful of the promise and perils of the existing and emerging technologies as they merge and converge to make the once unachievable imagination possible. With hundreds of participants and hundreds of thousands supporting our initiative, we are steadily making progress in creating the much-needed education and awareness of the critical security risks facing not only us individually or our nations but the very future of humanity. Watch/Listen Risk Roundup: Vodcast/Podcast to understand strategic security risks.

Risk Group Philosophy: Risk Group believes that risk management, security, and peace walk together hand in hand. Though security is related to managing threats and peace to managing conflict, risk management is associated with managing security vulnerabilities and managing conflict. It is not possible to conceive any one of the three without the existence of the other two. All three concepts feed into each other. Risk Group believes that the security we build for ourselves is precarious and uncertain until it is secured for everyone across nations. Tradition becomes our security, so if we create a culture of managing risks effectively, it will lead us to security, and security will lead us to peace.

Risk Group Thought Leadership: Risk Group's Collective Strategic Security Risk Analytics Platform is neutral, non-partisan, and non-biased towards any country, ideology, faith, religion, culture, or technology. Risk Group focuses on creating a strategic security thought leadership/scholarship to make our communities, nations, and the human race resilient to the rapidly emerging security risks from across quantumspace, cyberspace, aquaspace, geospace, and space (QCAGS). Risk Group appreciates the thought leaders from around the world who have stepped up to create thought leadership/scholarship to protect humanity's future. Please go to Risk Group and find out how you can contribute.

Risk Group Idea Incubator: Risk Group launches a new *Idea Incubator Platform* to protect humanity's future. As Risk Group identifies the strategic security risks facing our collective future, the search for new ideas, innovations, and inventions to manage these strategic security risks is at the top of the Risk Group agenda. Focusing on protecting humanity's future, Risk Group is embarking on a new collective idea incubation process to take the concept to commercialization. Risk Group's security-centric collective platform, thought leadership incubator, and idea incubator provide futuristic insights and tangible solutions. Contact Risk Group to understand what you can get from a thought leadership incubator.

Risk Group Membership: The risk group's focus is on defining the security-centric operating system for humanity's future. The development of a strategic security risk intelligence platform enables a collective effort towards shaping the future. Risk Group members will benefit from the strategic security risk intelligence platform, the growing thought leadership, credibility, influence, reach, and targeted focus to solve complex security problems facing humanity. Contact Risk Group to understand how to be a Risk Group member.

Risk Group Published Books

Stay tuned for Risk Group's Upcoming Books on Cyber Warfare, Artificial Intelligence, and Pandemics

**READ STRATEGIC SECURITY THOUGHT
LEADERSHIP ON RISK GROUP**

Introduction

The Risk Group's Strategic Security Risk Report 2021 is published as the world undergoes a deadly coronavirus COVID 19 pandemic and looks to 2021 for relief. At the core are the concerns about the human ecosystem showing visible signs of imbalance. Protecting planetary health to protect the human species dominates humanity's strategic security risks. Since everything is interconnected, understanding the why, how, when, and what of the pandemic requires evaluating the core of the human domains, dimensions, and ecosystem. It is, therefore, justifiably a focus of the 2021 Strategic Security Risk Report.

On the surface, the COVID-19 pandemic brings to light the vulnerabilities of human health to known/unknown pathogenic microorganisms. The on-going human health crisis seems no different than the previous pandemics (viral) crises we have experienced over the years. What is different is the advanced understanding, knowledge, and insights we have achieved over the years to the connected human ecosystem—all its domains and dimensions.

According to an article published in Science Alert, in April 2017, scientists discovered that octopuses and some squid and cuttlefish species consistently edit their RNA (ribonucleic acid) sequences to adapt to their environment.

There is no dispute that our environment and ecosystem are rapidly changing, and humans need to learn to adapt to a rapidly changing environment. So, the question is, how do we not only survive changing conditions but also how do we adapt and evolve.

While octopuses edit their own RNA sequences to adapt to their environment, how do humans access their RNA sequences to adjust? Is it possible for humans to access the necessary RNA strands from viral outbreaks such as the current COVID 19 pandemic?

An article published in the Journal of Microbiology, Immunology, and Infection (Science Direct, March 2020) reports COVID-19 to be a spherical or pleomorphic enveloped particle-containing single-stranded RNA associated with a nucleoprotein within a capsid composed of matrix protein. Is this RNA invasion in the form of the COVID 19 pandemic meant to adapt humans to get our species used to the changing environment?

Over the years, we have understood that humans are not autonomous organisms; we are superorganisms. Although we consider ourselves distinct, superior autonomous organisms, the growing understanding of microorganisms' role in human health is shaking autonomy fundamentals. The emerging understanding is that microorganisms are regulating human health. So, it is time to evaluate whether we understand the microbiopolitics centered around human health regulation.

The human genome is full of viruses. Viruses are powerful, ancient, and vital to our existence. We must build an understanding of our existence and evolution by evaluating the interconnectedness of living and non-living things. The tools and technology that we can access can help explain how we manage and eradicate current and future health crises. Progress and development in science and technology mean nothing if we do not apply emerging integrated knowledge to solve our species' most significant problems.

Our efforts should focus on understanding the drivers, evaluating the ecological imbalances, visualizing scenarios, and reporting the findings. Our research's overarching objective is to provide insight into the strategic security risks that could put the future of humanity at risk. Our actions are towards sustaining human life on Earth and beyond. For that, it is crucial to research, report, and manage the foundational imbalances along with their triggering factors.

Risk Group is determined to understand the interconnectedness of the living and non-living across each known domain and uncover the unknowns that impact our species. This yearly initiative is a step towards that.

Jayshree Pandya, Founder, and CEO, Risk Group LLC

Top Ten Strategic Security Risks

The Top 10 Strategic Security Risks Facing Humanity In 2021	
1	Planetary Health is In Peril
2	The Invisible Human Regulators
3	Vanishing Human Intelligence
4	Weapons of Mass Destruction
5	Gene Editing and Synthetic Biology
6	Self-Replicating Nano Machines
7	Artificial Super Intelligence
8	Super Volcanoes
9	Cosmic Threats
10	Unknowns-Unknowns

1. Planetary Health Is In Peril

As the human ecosystem shows visible signs of imbalance and deterioration, protecting the planetary health to ensure our species survival dominates this year's and future year's strategic security risks.

Need For Balanced Human Ecosystem

Risk Group recognizes that planetary health is in peril. Since human health and our planet's health are inextricably linked, our species' survival depends on a balanced human ecosystem in all domains and dimensions.

Human life and the life of all living beings depend on a sensitive balance in our planetary ecosystem. Research since the earliest of civilization has discovered these foundational truths: all living beings require water, nutrition, and food for survival. Most living organisms also need oxygen. So, it is essential to understand and evaluate:

- What processes link the living (humans, microorganisms, and more) to the nonliving world?
- What are the everyday needs and resources for human's survival and security?

Planetary Processes

All living beings on our planet have a role to play and have a distinct purpose. Beyond these life forms, non-living elements on our planet also play a role—just like cosmic elements in this universe. Our current understanding of the interplay between living and nonliving beings has barely scratched the surface, especially the unknowns around nonliving structures. This emerging understanding takes us on a new journey of reevaluating the concept of "living."

According to a paper titled BIZARRE SUBATOMIC "QUASIPARTICLE" REPRODUCES LIKE A LIVING CELL, scientists have found that a bizarre subatomic "quasiparticle," called a skyrmion, is capable of reproducing itself unusually—they actually split in two and reproduce.[1]

There is a need for us to understand more. We need an elevated understanding of which planetary processes drive human existence. We require a sophisticated awareness of how the planetary processes influence and connect the living and nonliving forms.

At the core, there is no dispute that Earth's natural systems—the air, the water, the biodiversity, the climate—are our life support systems. There is also no dispute that there is a constant energy flow between the human ecosystem and its domains and dimensions. It is also undisputed that the human ecosystem's energy flow begins in the form of light energy, which is further transformed into chemical energy and then heat energy. While many cellular processes support energy transformation, it is crucial to understand that solar energy is and will remain at the center of the human ecosystem.[2]

The human ecosystem also requires elements such as carbon, nitrogen, or phosphorus from the planetary ecosystem. These elements are continually cycled between the living and the non-living state, and each of these processes is interconnected and complex. So, how should we study these processes from the perspective of human existence and planetary health?

BIZARRE SUBATOMIC "QUASIPARTICLE" REPRODUCES LIKE A LIVING CELL-Futurism https://futurism.com/the-byte/bizarre-subatomic-quasiparticle-splits-like-living-cell?
Is there a Need For an Ecological View Of the COVID 19 pandemic? Risk Group. https://riskgroupllc.com/is-there-a-need-for-an-ecological-view-of-the-covid-19-pandemic/

It seems that scientists have been studying biogeochemistry, geology, and chemistry of the Earth using the existing principles and tools. Global warming and environmental pollution have been evaluated by studying elemental ratios, mass balance, and element cycling. The elemental ratio **C: N: P: Fe** determines the expected ratio of living organisms' nutrient elements. The question is whether this ratio will evolve our understanding of the on-going changes and their implications on the entire human ecosystem. The goal is to understand the biogeochemical cycles that determine the human ecosystem.[3]

Scientists also use the mass balance equation to describe the state of a system. The question is whether applying the mass balance equation where NET CHANGE = INPUT + OUTPUT + INTERNAL CHANGE will indicate the changes in the human ecosystem that cause imbalances. Suppose we use this equation to understand the shifting health of the planetary ecosystem; in this equation, the net change in the human ecosystem from one time period to another will be determined by the variable inputs, outputs, and internal changes. So, the question is:

- How do we define and quantify these inputs?
- How do we define and quantify these outputs?
- What internal changes are in the scope of this evaluation?

We must understand the drivers of the energy and nutrient flow to effectively protect the planet, considering it is tied to human sustainability. [4]

Energy Flows

Energy flows through the human ecosystem and drives the nutrient flow. The matter is recycled. In a human ecosystem, the energy always flows from the Sun to the producers (plants) and consumers and eventually to decomposers that return the soil's energy. Humans are consumers, and microorganisms usually fill the role of decomposers.

---

[3] Biogeochemistry-Science Learning Hub.
https://www.sciencelearn.org.nz/resources/958-biogeochemistry

[4] The Ecosystem and How It Relates To Sustainability- Univ. Of Michigan.
https://globalchange.umich.edu/globalchange1/current/lectures/kling/ecosystem/ecosystem.html

Energy flows through the human ecosystem, usually in the form of heat. Basic chemical elements are recycled. Since the human ecosystem is made of living and nonliving things, it is essential to evaluate the living and nonliving linkages and understand the rapidly evolving energy transformation that impacts the biogeochemical cycle. We have not invested enough resources in understanding these linkages spanning the entire human ecosystem in all its domains and dimensions. Nor have we focused on identifying the interfaces of the human/animal to living/nonliving.

Diverse factors are responsible for the balance of the human ecosystem and the health of our planet. Each human action at the micro and macro level directly relates to the balance of our ecosystem. The pathways of human ecology have not yet been analyzed comprehensively and quantitatively. The increasing imbalance created in the human-microbial-animal-ecosystem interface likely directly correlates with the human-physical ecosystem.

Research into environmental factors like changes in climate, domains, dimensions, habitat, etc., along with human host and microbial factors, has advanced in recent years. But there is a need for a deeper understanding. Careful and thoughtful analysis of our habitat is required to avoid species extinctions and ensure future human existence because of the interdependencies of a properly functioning ecosphere.

The Growing Imbalances

Human life on Earth might have to confront the worst imbalances that we have faced yet due to the expected ecological backlashes created by natural and human-made destruction. Since the environment is complex and without a comprehensive understanding of the root causes, environmental regulations are challenging to implement.

Surveillance systems with real-time tracking, warning, and response capabilities will give us a much-needed understanding of the existing and upcoming environmental risks. Identifying the risk catalysts is critical to prevent, predict, and contain the growing imbalances.

Since the human ecosystem is an essential component of human existence, it is fundamental to ensure its sustainability. We must confront changes to our ecology. Therefore, a need exists for a collective effort, initiatives, and investment to ensure we use our species' collective intelligence and tools to protect our planet. It is time to apply collective resources to save our ecosystem.

Risk Group Efforts

Risk Group works towards formulating explanatory frameworks that could address the growing imbalances in our planetary health and why it matters for human survival and security. It will perhaps begin a conversation for a new approach towards protecting our planet.

2. The Invisible Human Regulators

Although humans consider themselves distinct, superior autonomous organisms, the growing understanding of microorganisms' role in human health and well-being is shaking autonomy fundamentals.

The Human Organism

The growing understanding of microorganisms' role in our health and well-being is jolting the foundational belief that humans are autonomous beings. The emerging understanding is that microorganisms are regulating humans. So, it is time we evaluate whether we understand the microbiopolitics centered on the regulation of humans. [5]

Each human has a unique microbial signature cloud. The emerging research reveals that the low diversity of microorganisms in humans correlates with both communicable and non-communicable diseases. From allergies to arthritis, diabetes to cardiovascular diseases, obesity to psychiatric disorders, inflammatory bowel diseases to intestinal disorders, microorganisms play a vital role in human health and disease. [6]

Exposure to diverse microorganisms and establishing proper symbiotic relationships is vital for the human immune system to function optimally. It is the exposure to necessary microorganisms' diversity that perhaps trains and influences our immune system to induce adaptive immunity and initiates memory (B and T cells) fundamental for our defense.

---

[5] The Microbiome-How Gut Bacteria Regulate Our Health-New Scientist. https://www.newscientist.com/article/2254471-the-microbiome-how-gut-bacteria-regulate-our-health/

[6] The impact of human activities and lifestyles on the interlinked microbiota and health of humans and of ecosystems-ScienceDirect. https://www.sciencedirect.com/science/article/pii/S0048969718303413

The human immune system's adaptability is essential for defending us from infectious diseases (communicable) and developing non-communicable diseases. Thus, understanding our microbiome's basics and its role in regulating our health, our adaptability, and evolution is becoming crucial for our very survival as a species.

Human Health and Microorganisms

Humans have co-evolved with the microorganisms and have a unique set of microbiotas in our ecosystem and environment. The microbiome is the human ecological community. This ecological community comprises constructive and destructive elements – symbiotic and cooperative, to pathogenic microorganisms that make and share our bodies. Humans are the host to this large unique community, and together we make us humans.

It is believed that the number of microorganisms that colonize human host cells is ten times in number than the host cells. Any changes in these diverse microbial communities unique to each one of us may be responsible for most of our communicable and non-communicable diseases. Despite the importance, these microbial communities residing within us remain mostly unstudied. There is a need for a concerted research effort to understand this relationship better.

Advancing our understanding will be the basis of a new way of approaching human health and disease and finding new ways to prevent and respond to diseases, irrespective of whether the disease is contagious or non-contagious.

Human Microbial Community

Humans are used to living in communities made of diverse humans and microorganisms. Our living communities, micro, and macro are slowly but surely evolving towards the interconnected world we see today.

It seems the microorganisms in us, on us, and around us mostly follow this cooperative mode of living. Of the numerous characteristics that humans possess, many-core attributes are seen in microbial life as well. For example, besides having a diversity of form and structure, microorganisms also form conglomerates or communities.

In the human health ecosystem, microorganisms seem to have a promising role to play. As a result, there is a need to further evaluate microorganisms for their impact on the human body and its metabolism and how they do it. It will help us understand how microorganisms and the human host can best work together, their ecological relations, and evolve this relationship to be productive for human health and wellness, adaptation and evolution, sustainability and security. And for that, the human ecosystem needs to be seen through the lens of microbiology.

To understand how changes in normal microbial populations affect or are affected by the human environment, we first have to establish what normal is and if it is achievable in our rapidly evolving human ecosystem.[7]

A Cooperative Ecosystem

As we work towards defining the norms, it is crucial to understand why, how, and when living structures in our ecosystem interact. The human body is a collaborative ecosystem where trillions of microorganisms live in and around it. The collective community of these known/unknown microorganisms consists of bacteria, viruses, fungi, and more. It is essential to understand the nature, mode, and impact of this human-microbial relationship and evaluate:

- The microorganisms that are part of the human ecosystem
- Their role in human health and metabolism
- The drivers of the changing microbial flora in the human ecosystem
- The role microorganisms play in human adaptation to the changing environment

There is also a need for a deeper understanding of the human ecosystem and its impact on the microbes that live in, on, and around the human body.

Risk Group Efforts

---

[7] Humans Have Ten Times More Bacteria Than Human Cells-Science Daily. https://www.sciencedaily.com/releases/2008/06/080603085914.htm

Risk Group works towards formulating explanatory frameworks that could adequately make sense of the human-microbial relationship in the rapidly changing human environment and why it matters for our adaptation and evolution. It will perhaps begin a conversation for new kinds of attitudes and approaches towards micro to macro integration and ecosystems for human survival and security.

3. Vanishing Human Intelligence

While conflicting views are surrounding declining human intelligence, it is crucial to evaluate whether humans' genetic makeup has led to this steady degeneration and if the technological advances are being a catalyst in this change.

Shrinking Human Brains

Homo sapiens had larger brains in relative size approximately 30,000 years ago. There are conflicting views about shrinking human brains. There is research that human brains are decreasing in size over the years and that human brain size is correlated with intelligence (though it only explains 2% variation).

While scientists are still working to understand the implications of shrinking intelligence, the implications are challenging for humanity's future. Amidst the decreasing size of the organ of the human intellect, reports are also emerging of deliberate efforts to augment human intelligence with artificial intelligence. [8]

There is a prediction that humans will become hybrids (merge with artificial intelligence) in the next decade. That indicates our brains will be able to connect directly to the cloud, where there will be thousands of digital computers, and those computers with enormous digital data will augment our existing human intelligence with machine intelligence. [9]

---

[8] If Modern Humans Are So Smart, Why Are Our Brains Shrinking? Discover Magazine. https://www.discovermagazine.com/the-sciences/if-modern-humans-are-so-smart-why-are-our-brains-shrinking

[9] Ray Kurzweil: "AI Will Not Displace Humans, It's Going to Enhance Us"- FUTURISM.
https://futurism.com/ray-kurzweil-ai-displace-humans-going-enhance

It is also possible that the brain will connect via nanobots –
which are tiny robots made from DNA strands. Irrespective of
how our brains connect to digital machines, our thinking will be
a hybrid of biological and non-biological (artificial) thinking in
the coming years.[10]

The more advanced the digital data cloud, the more evolved our
thinking will be based on information and intelligence. So, if the
technological planning works as has been projected in the next
decade or two, our thinking will be predominantly non-
biological. That brings us to an important matter: What will be
the implications on the human intelligence operating system?

Declining Human Intelligence

There are visible signs that human intelligence seems to be in
decline. It is not only the IQ that is reported to be declining, but
the emotional stability is also in fall. The reasons are both natural
and human-caused. It is not only the genetic makeup of humans
but also technological advances that are also causing the decline.
We use technology to solve problems, and we use technology to
make every day-to-day decision essentially. Amid that, and with
our existing conditions, the questions we need to ask ourselves
are:

- Can we remain intellectually alive?
- Can we have a clear-sighted view of our security risks?
- Can we have a broad range of ideas to solve the problems
  in front of us?
- Can we have the emotional stability to focus on the
  solutions?
- Can we separate ourselves from the paternalistic
  treatment coming our way in the form of artificial
  intelligence that will be fed to us?
- Can we resist the mass manipulation happening across
  social (media) in the form of human misinformation and
  disinformation?

---

[10] Ray Kurzweil: Humans will be hybrids by 2030.
https://money.cnn.com/2015/06/03/technology/ray-kurzweil-
predictions/index.html

When we supplant how our brain's exercise with the dependency of digital screens, there is a need to acknowledge that human intelligence's foundations are susceptible to compromise. Since the human mind is similar to a muscle, the more we use it, the more functional it will remain. Once we stop using it, it will undoubtedly decline.

The concerns about the declining human intellect—the ability to think for ourselves—blindly following a few, trusting all machine intelligence, and allowing these influences to shape our collective thinking poses an enormous risk for humanity's future. It is crucial to evaluate whether we should be training our minds to think more independently. It is also vital that we assess the implications of being susceptible to herd mentality. Furthermore, the question remains whether we need to focus our resources on human enhancement versus developing artificial intelligence. Also, should we reorient our focus to the research and development of the human mind serving as a primary processor?

The current trends of defaulting human minds to technology, allowing other intelligent algorithms to drive decision-making, and replacing vital mental processes will have serious implications for our species' security.

We should not be sacrificing the minds of the masses in the name of technological progress. When analytical thinking and innovation are needed for our species' survival skills, manipulating minds with data and decisions fed by another intelligent species, inherently programmable, is a mistake.

Artificial Intelligence (AI) is simply a tool and needs to remain a tool that complements our necessities. As we put more effort into advancing our tools, we need distinctly to evolve and advance our human capabilities in parallel with machine capabilities.

Risk Group Efforts

Risk Group works towards formulating human enhancement frameworks that would effectively advance our human capabilities to compete with rapidly growing artificial intelligence.

The goal is to begin a conversation for new kinds of attitudes and approaches to change our blind reliance on artificial intelligence and refocus that R&D on enhancing human intelligence for our survival and security.

4. Growing List of Weapons of Mass Destruction

While wars are never really won in their real sense, humans continue to develop new weapons of mass destruction.

War and Destruction

The world has already seen two World Wars. While we have seen war and destruction since the beginning of civilization, weaponry innovation and invention continues. According to the Council on Strategic Risks, there are worrisome signs that weapons of mass destruction (WMD) are reemerging with a purpose to increasingly threaten international stability and put the future of humanity at risk.[11]

Over the years, only nations/states have had the capacity and resources to develop traditional WMDs. The reason is these weapons required significant capital, infrastructure, and intellectual capacity to build and maintain. However, in recent years, emerging technologies have changed this paradigm.

The trends are shifting from nations/state players to small groups and even individuals. The democratization of destruction has propelled the weaponization of emerging technologies. The reduced financial, intellectual, and material barriers required for WMD development has played a role in its commercialization.

Weapons of Mass Destruction

Technological change is a prominent contributor to weapons of mass destruction as seen historically and in the present reality. Emerging technologies will continue to enable the rise of new, agile threats to our collective security.

---

[11] Weapons of Mass Destruction-COUNCIL ON STRATEGIC RISKS. https://councilonstrategicrisks.org/wp-content/uploads/2019/10/Weapons-of-Mass-Destruction_The-State-of-Global-Governance_2019_10_20.pdf

The growing list of weapons of mass destruction (WMD) — beyond nuclear, chemical, and biological weapons, will increasingly capitalize on human domains' asymmetric destructive capabilities.

The weapons of mass destruction that humans have developed and continue to establish keeps growing, and there is no end in sight. For instance:

- Synthetic Biology: Technological changes in the biological sciences are incredibly rapid. It includes advances in synthetic biology and gene editing. The use of sophisticated gene-editing techniques and tools to sequence, synthesize, and manipulate genetic material is an emerging threat for mass destruction weapons. Moreover, genome-sequencing technologies read DNA sequences and convert them into digital information. In contrast, gene synthesis technologies fundamentally do the contrary, allowing scientists to translate digitized genomic data from a computer into physical DNA sequences. When scientists can modify or recreate living organisms in a lab environment using these sequences, it is a cause of great concern as any pathogen can be created anywhere in the world, putting our collective security at risk. [12]

- Additive Manufacturing: The proliferation of digital information due to additive manufacturing is growing the risks of WMDs. Additive Manufacturing (AM) is a rapidly emerging technology that includes several technologies, including 3D Printing, Rapid Prototyping (RP), Direct Digital Manufacturing (DDM), layered manufacturing, and additive fabrication. The availability of information and advances in additive manufacturing stands to change the power of long-held Government export controls that limited the materials and knowledge needed in WMD programs. Since it is easier to produce

---

[12] DOD Officials discuss countering WMD, Threat posed by Synthetic Biology-US DOD. https://www.defense.gov/Explore/News/Article/Article/1128356/dod-officials-discuss-countering-wmd-threats-posed-by-synthetic-biology/

WMDs using additive manufacturing covertly, it is a cause of great concern for war and peace. [13]

- Unmanned Aerial Systems: Robotics and machine learning are enabling more rapid and cheaper bioproduction. The explosion in data these advances are driving, and other trends that direct the characteristics and capabilities of semi-autonomous or autonomous unmanned aerial vehicles (UAV) that carry weapons of mass destruction (WMD) is a growing concern.[14]

- Semi-Autonomous to Fully Autonomous Systems: Lethal autonomous weapons systems that can identify, select, and engage a target without meaningful human control, which can be highly scalable, represent a new category of weapons of mass destruction. [15]

- Nanomachines: Microscopic machines can perform functions at the cellular, molecular, and atomic levels to execute the processes of self-replication, self-repair, and self-assembly. They could theoretically also alter environments, structures, and living beings rapidly from within drastically. The potential of such machines could lead to a *doomsday scenario*. The reasoning is that nanomachines can present with a self-replicating behavior, and *nanodevices* can be destructively used against, for example, the atmosphere or oceans. It can become uncontrollable, exponential, and if and when it becomes a reality, it has the power to destroy humanity. [16]

---

[13] 3D printers could help spread weapons of mass destruction- Scientific American. https://www.scientificamerican.com/article/3-d-printers-could-help-spread-weapons-of-mass-destruction/

[14] Unmanned Aircraft Systems-CISA. https://www.cisa.gov/sites/default/files/publications/uas-ci-challenges-fact-sheet-508.pdf

[15] Lethal Autonomous Weapons System-Future of Life Institute. https://futureoflife.org/lethal-autonomous-weapons-systems/

[16] Study Finds Self-Replicating nanomachines feasible-Foresight Institute. https://foresight.org/study_finds_self-replicating_nanomachines_feasible/

- Biological Weapons: Biological weapons that use biological organisms like a virus to create a disease can affect millions of people and could prove to be uncontrollable once released, becoming an existential security risk. [17]
- Chemical Weapons: Chemical weapons make use of toxic chemical products, causing blisters, bleeding, asphyxiation, or nerve agent responses that are also hard to control. [18]
- Fission Bomb: Fission bombs, also called atomic bombs, get their power purely from fission reactions. The boosted fission weapon is a fission bomb that uses fusion reactions to boost its yield. [19]
- Fusion Bomb: Also called the hydrogen bomb, fusion bombs rely on fusion reactions between isotopes of hydrogen called tritium and deuterium. [20]
- Neutron Bomb: The weaponized Neutron Bomb is detonated with a blast of neutron radiation—and is a salted bomb, in which cobalt or gold or some other suitable materials surround the weapon. The resulting explosion will yield a substantial amount of radioactive contamination. [21]
- Napalm Bomb: This bomb is made of petroleum jelly and was used extensively as an incendiary bomb in the Vietnam War and was subsequently banned in military

---

[17] Biological Weapons-John Hopkins Center For Public Health Awareness. https://www.jhsph.edu/research/centers-and-institutes/johns-hopkins-center-for-public-health-preparedness/tips/topics/Biologic_Weapons/BioWeapons.html

[18] Weapons of Mass Destruction-Department of Homeland Security. https://www.dhs.gov/topic/weapons-mass-destruction

[19] The Manhattan Project-US Dept of Energy. https://www.osti.gov/opennet/manhattan-project-history/Events/1890s-1939/discovery_fission.htm

[20] Hydrogen Bomb-Atomic Heritage Foundation. https://www.atomicheritage.org/history/hydrogen-bomb-1950

[21] The Neutron Bomb-Air Force Magazine. https://www.airforcemag.com/article/The-Neutron-Bomb/

conflict by the 1980 United Nations Convention on Certain Conventional Weapons (CCW).[22]

- Radiological Weapons: involves the dispersion of radioactive materials by using an explosive device and is considered a medium level weapon of mass destruction.
- Toxicological Weapons: employ various natural and artificial poisons and are highly effective in causing human death.

World War I brought us widespread use of chemical weapons. The world has also seen the devastation of nuclear weapons. Weapons of mass destruction remain a daunting international security challenge.

These threats have become more complex, and in recent years norms and international controls regarding WMD possession and use have been weakening. Since these trends in weapons of mass destruction require new and expanded efforts to strengthen our independent and collective security, it is crucial to evaluate how to permanently rid the existing and emerging world of weapons of mass destruction.

Risk Group Efforts

Risk Group is working towards identifying emerging weapons of mass destruction driven by the democratization of destruction enabled by the making of cyberspace. We hope that this reporting will perhaps begin a conversation for new kinds of attitudes and approaches towards emerging technologies' governance.

---

[22] Liquid Fire-How Napalm was used in the Vietnam War-War History Online. https://www.warhistoryonline.com/vietnam-war/history-napalm-vietnam-war.html

5. Gene Editing and Synthetic Biology

No scientific achievement has brought as much possibility and potential for transforming the human ecosystem and vulnerability to deliberate abuse and misuse as Synthetic Biology.

Beyond Natural Systems

Today, we, the humans, are no longer confined to nature. The advances in science and technology have given us the capability to go beyond these natural boundaries and create *living things*. While these boundaries remain the foundation to our natural biological ecosystem, we, the humans, are now capable of building on augmenting this foundation and creating a human-made, synthetic bio-ecosystem of our own desire and design. We have reached a point where tools and technology to define and design entire human and non-human genomes have caught up with our ability to construct them.

In the coming years, we will be able to define, design, and construct any cell, organism, or any biological species we want—with a growing selection of genomes from human, non-human, and or any other living species. Scientific development is almost at the point where to create or manipulate *life* will only be constrained by our imagination. Soon, we will be able to harness the power to build any cell, organism, or biological species up from scratch. [23] [24]

---

[23] Synthetic Biology-ScienceDirect.
https://www.sciencedirect.com/topics/engineering/synthetic-biology
[24] Self-organizing robot armies produced - RT.
https://www.rt.com/news/self-organizing-termite-robots-172/

There is a growing belief that there is no such thing as an *artificial gene* and that, in fact, what matters to a gene is its DNA sequence and not how one makes that DNA sequence. This ability to synthesize DNA sequences from living and non-living species have completely transformed this domain.

Synthetic Biology

Synthetic Biology integrates computational analysis, biological data, and the systems engineering paradigm to design new synthetic biological machines and systems. The recent advances have enabled scientists to make new sequences of DNA from scratch. By merging these advances with modern engineering principles, anyone can now use computers and laboratory chemicals to design organisms to meet the desired objective. [25]

Humans have been altering plants and animals' genetic code for ages by selectively breeding individuals with desirable features. Whether it be a new cell, organism, or species, by combining elements of engineering, chemistry, computer science, and molecular biology, synthetic biology now focuses on writing and programming new DNA to create genetic machines from scratch. [26]

As we have evolved our ability to read and manipulate genetic code, we have begun to take specific genetic information from one organism and add it to another to achieve the desired result. Since we are on the verge of creating new cells, or new microorganisms, plants, animals, and perhaps humans in the coming years, the question is whether we are redefining the very meaning of evolution.

Since there is a growing concern around the idea of constructing new forms of life, it is crucial to evaluate the ethical considerations of whether synthetic life experiments are an attempt to probe the origins of life. [27]

---

[25] The challenges of informatics in synthetic biology: from biomolecular networks to artificial organisms-NCBI.
https://www.ncbi.nlm.nih.gov/pmc/articles/PMC2810114/

[26] Genetically Modified Organisms-National Geographic.
https://www.nationalgeographic.org/encyclopedia/genetically-modified-organisms/

[27] Synthetic Biology: Origin, Scope and Ethics -CENTRE FOR HUMANS AND

Technology has now allowed us to complement evolution with human-made technology. It will enable us further to decode a genome and break it down to its specific DNA sequences the way one would decomp the code of a software program. We can then design whatever we want and recompile it anywhere. At that point, we can even make disposable biological systems eliminating the role of natural reproduction. We can make much simpler organisms. Although we speak about the biological future that can be engineered, it is crucial to evaluate whether we have the formidable knowledge of preventing unexpected security risks. [28]

Security Risks

What are the potential security risks? In addition to several scientific and technical challenges, synthetic biology raises many critical biosecurity issues that can put the future of humanity at risk. In the absence of integrated risk analysis, managing security risk is difficult to achieve. Beyond regulating DNA synthesis companies, the fundamental problems of security surrounding amateur biologists' biohacking efforts remains a source of grave concern. Moreover, the distributed nature of open-source biotechnology and free-market accessibility makes it challenging to track potential biosafety and biosecurity problems.

There is a need to evaluate:
- How will we control unexpected complexities?
- How will we track or manage biosafety and biosecurity risks?
- How will we control the accidental release of an unintentionally harmful organism or system?
- How will we control the purposeful design and release of an intentionally harmful organism or system?
- Is it safe to manipulate and create life?
- How likely are accidents that could unleash organisms onto a world that is not prepared for them?

NATURE. https://www.humansandnature.org/synthetic-biology-origin-scope-and-ethics

[28] A Life of its Own-New Yorker.
https://www.newyorker.com/magazine/2009/09/28/a-life-of-its-own

- What steps can be taken to mitigate these challenges?

Synthetic biology brings to light several questions, including who will have control and access to synthetic biology products and innovations? These are reasonable concerns and questions. Assessing these risks in the real world is complicated as we do not know how these synthetic organisms will interact with pollinators, biological and non-biological systems in the human ecosystem.

Risk Group Efforts

Risk Group works towards formulating a security risk management framework that could advance our capabilities to identify emerging synthetic biology security risks proactively. It will perhaps begin a conversation for new kinds of attitudes and approaches towards a need for a practical governance framework.

6. Self-Replicating Nanomachines

As we get closer to building nanomachines that will be able to rearrange atoms to construct new objects, it is crucial to evaluate where self-replicating nanomachines can take humanity and whether we will be able to put rails around its development.

Nanomachines

Nanomachines are tiny devices that are built from individual atoms, and their size is measured in nanometers. There is an intense effort towards designing nanomachines that can self-replicate. When and if this objective is achieved, anything produced by nanomachines will be transformative. It will also be enormously inexpensive because the technology, once achieved, will be self-replicating. It will not require specific materials—which might otherwise be rare and cost money. If we manage to build nanomachines that can rearrange atoms, a world of exciting possibilities will open up. [29]

It is believed that nanomachines could be used to:
- provide innovative treatments for many communicable and non-communicable diseases
- provide quick and effective treatment for all types of cancer
- repair damaged tissue and bones
- support bones and muscle tissue by constructing molecular support structures by reassembling nearby tissue
- turn any material into food
- build better products to satisfy the demands of our growing population

---

[29] A nanometer is about 1 billionth of a meter

- solve many environmental problems—such as ozone depletion and global warming

The advances in nanotechnology and the development of self-replicating nanomachines could signal a new age for humanity's future. It will take us to an era where food scarcity, illness, and environmental problems could be solvable. With the emerging ability to manipulate human cells at the atomic level, medical science will rapidly devise treatments for most human illnesses. And since nanomachines can be designed to make copies of themselves, these treatments will be inexpensive and accessible.

Self-Replicating Machines

Nanomachines, devices built from individual atoms, are close to being a reality. In the coming years, nanomachines will be injected into living cells to fight disease and repair cells. It will also help nations to literally convert soil into food, eliminate poverty, and much more. As we await the future, it is crucial to evaluate whether it will be possible:

- to produce machines the size of atoms
- to build nanomachines that can evolve objects from the atom up
- for nanomachines to create copies of themselves
- As efforts are focused on developing such machines, it is crucial to evaluate problems that can surface when exploring self-replicating machines. For instance:
- Will these nanomachines be controllable?
- If their reproduction escalates exponentially, will it put our entire planet in danger?

Undoubtedly, nanomachines offer humanity hope for the future. While progress and development should continue, unless we have effective ways to control the dangers of producing self-replicating machines, the security risks perhaps outweigh the potential gains. For this reason, unless we figure out how to manage these self-replicating machines, we should not be building them commercially.[30]

---

[30] Nanotechnology pioneer slays 'grey goo' myths-Science X.
https://phys.org/news/2004-06-nanotechnology-grey-goo-myths.html

Complex Challenges

The weaponization of nanomachines and the inevitable use of this technology in warfare is a growing risk. Self-replicating nanomachines that can be designed to target and destroy organic material could be released over enemy territory. It could reduce the enemy territory population to dust within a matter of hours. [31]

The most significant risk around nanomachine technology involves the self-replicating feature. Imagine that a nanomachine can make a copy of itself by rearranging the atoms contained in any nearby matter. Since it produces a replica of itself, the *offspring* machine will also likely be able to replicate. After a few generations of replication, we would have millions of nanomachines on our hands. So, without a way to control nanomachines' reproduction, the planet will be in danger of being overrun by nanomachines.

Furthermore, since the nanomachines will clearly be using the planet's resources as raw material to replicate, the threat is that the planet could lose vital resources and transformed into a seething mass of nanomachines. Since molecular machines can replicate exponentially and transform the planet, it is crucial to evaluate the existential risks to humanity's future.

Risk Group Efforts

If self-replicating nanomachines spiral out of control, they could alter our planet to such an extent that it will no longer be suitable for biological life. As a result, we must be cautious about our survival. Risk Group is working towards understanding existential risks and whether there is any possibility in controlling self-replicating machines.

---

[31] Nanoweapons: A Growing Threat to Humanity by Louis A. Del Monte

7. Artificial Superintelligence

As the intelligence explosion seems inevitable, Artificial Super Intelligence's troubling trajectory forces us to think seriously about our objectives as a species.

Intelligence Evolution

As humanity stands on the brink of digital transformation, the scale, scope, and intricacy of the impact of machine intelligence evolution go beyond anything humanity has faced before. Consequently, the speed at which the ideas, innovations, and inventions are emerging on the back of artificial intelligence has no historical precedence and is fundamentally disrupting everything in the human ecosystem as we know it. [32]

Also, the breadth, depth, and impact of the intelligence evolution on furthering the ideas, innovations, and implementations across the human domains indicate the changing times.

The technology-triggered machine intelligence evolution and the linkages between ideas, innovations, and trends have brought us to the doorstep of *superintelligence*. Irrespective of whether we believe that *superintelligence* will take shape, the very thought raises many concerns and brings critical security risks to the forefront. It compels us to begin a dialogue on our priorities as a civilization.

Although there is no way to calculate just how and when this *superintelligence* machine evolution will unroll, one thing is clear: it changes the fundamentals and foundation of our collective security. The response to it, therefore, must be integrated and comprehensive for the future of humanity.

Information Age

---

[32] Artificial General Intelligence for Robotic Colonization-Risk Group. https://riskgroupllc.com/artificial-general-intelligence-for-robotic-colonization/

While human intelligence has changed our ecosystem numerous times, intelligent machines are expected to change and expand the human ecosystem in totally different ways in the coming years. As seen today, the connected computers, information, communication, digitization technology, and the internet has fundamentally disrupted the information process giving rise to an information age.

The success of the rapidly progressing information age is due to digital data. The data and information envelop all human domains. Upon evaluating the intelligent machine evolution, we realize that it has transformed the human ecosystem through structural innovations. Moreover, the quantity and quality of digital data and information created and stored by humans rapidly multiply in cyberspace. It has reached about several quintillion bytes of data already. When the current human population is around 7.7 billion, and we continue to be digitally interconnected, the digital data volume is expected to grow further. And with the exponential data growth, machine intelligence will continue to evolve.

Progress in Machine Learning Algorithms and Neuromorphic Chips

On the back of the information revolution, as the machine learning algorithms improve rapidly, the power and promise of self-learning software seem immense. Along with the formidable advances in software development, there also appears to be parallel evolution in computer hardware to enhance machine intelligence capabilities. It is clearly evident in the pace and investment towards developing systems on a chip.

The chips are focused on designing a more efficient, lower energy microprocessor chip that mimics the brain circuitry. As these *neuromorphic chips* are being designed to process sensory data such as images and sound and counter to changes in that data in ways not precisely programmed, a lot is expected to change for machine intelligence and intelligence evolution.

Primarily, minimizing the input power of a neural net onto a single semiconductor chip means that these learning and pattern recognition algorithms and technologies can be embedded into a broader range of systems —thereby increasing data and information for exponential growth in machine intelligence.[33]

As the rapidly evolving computers' computing power will exceed that of even the most developed human brain, machine intelligence's exponential progression will continue, and artificial superintelligence seems to be just around the corner.

Impact of the Intelligence Explosion

There is accelerating progress in both hardware and software. The impact of the artificial intelligence boom across all domains is an indicator of the approaching superintelligence and the inevitable impact on the human ecosystem. *It brings us to a fundamental question: what happens to human intelligence and the human race when the two grapple for dominance?*

While there is no doubt that when an artificial superintelligence emerges, it will bring to bear more sophisticated skills than the average human. *But would that not also mean creating another intelligent species that may or may not behave with human interests at its core?*

Keeping up with Super Intelligence

Amid the rapidly evolving, converging technologies, when the artificial superintelligence explosion seems inevitable, how will humans compete with super-intelligent machines?

It is believed that to overpower artificial superintelligence, the enhancement of human intelligence should be targeted. It is theoretically possible to create a "superhuman" with super intelligence to keep AI in check.

---

[33] The Rise of the Silicon Brain-Risk Group. https://riskgroupllc.com/the-rise-of-the-silicon-brain/

While in theory, creating a superhuman with superintelligence could happen through intelligence amplification and augmentation of the human brain (through advances in bioengineering, genetic engineering, nootropic drugs, mind uploading, and even direct brain-computer interfaces, AI assistants, and more), the reality is the human brain is going to be dependent on AI for the foreseeable future.

There is no clear timeline or agreement on when superintelligence is likely to be attained. But one thing is clear: the troubling trajectory of artificial superintelligence should lead us to reevaluate our species' priorities.

Security Risks

Since there is no direct evolutionary incentive for an AI to be friendly to humans, the challenge is in evaluating whether the artificial superintelligence will default to outcompeting humanity. The reality remains that the artificial superintelligence evolution will have no inherent tendency to pursue outcomes that align to human survival and prosperity— and there is hardly a reason to expect the contrary.

We, humans, live in a paradox as AI advances are shaping the human ecosystem that is both more perilous and plentiful than ever before. Whether promise or peril triumphs will be a defining force in humanity's future.

We need to begin discussing the troubling trajectory of artificial superintelligence evolution for humanity's future. Whether we believe that artificial superintelligence is near or here, the very thought raises crucial security questions for humanity's future, forcing us to think seriously about what we want as a species. If artificial superintelligence spirals, it could alter human existence; therefore, we must be cautious.[34]

Risk Group Efforts

---

[34] Protecting Humanity in the face of Artificial Intelligence-Risk Group. https://riskgroupllc.com/protecting-humanity-in-the-face-of-artificial-intelligence/

We must be cautious about our survival. There is a justifiable need to evaluate this risk further. Risk Group is working towards understanding the existential risks and whether a framework exists to control artificial superintelligence.

8. The Supervolcanoes

Could a giant Supervolcano end human life on Earth as we know it?

Volcano Eruption

Numerous volcanoes continue to erupt. It is crucial to understand that not all volcanic eruptions are a cause of concern for humanity's future. But the supervolcano in Yellowstone National Park could cause an ultra-catastrophe, which cannot be ignored. While the Yellowstone volcano in Wyoming does not erupt often, it is hard to know when it will erupt in totality again. It could happen anytime. If and when it happens, it is believed that the blast could kill billions, make the United States uninhabitable, and also could wipe out much of human life on this planet.[35]

Implications

There are many profound implications. The reason is:

- A supervolcanic eruption would propel miles of rock, dust, and ash into the atmosphere.
- It would create a flow of magma that would bury miles of area.
- The aerosols released could spread globally.
- The release of dust and toxic gases (like sulfur monoxide into the atmosphere) would create an acidic veil worldwide, reflecting sunlight.
- As the toxic cloud blocks sunlight, global average temperatures could plunge significantly — and not return to normal for several years. It would lead to a dramatic cooling of the global climate that would last at least numerous years.

---

[35] What is a Supervolcano? USGS. https://www.usgs.gov/faqs/what-a-supervolcano-what-a-supereruption?qt-news_science_products=0#qt-news_science_products

- Rainfall could decline sharply. That itself might be enough to trigger a die-off of tropical rainforests. It will also result in devastating crops. Farming/Agriculture could collapse. Without sunlight, warmth, and food, a significant loss of life would be incurred.
- It will also destroy the power grid.

The implications are not just for the United States. It could be the greatest catastrophe since the dawn of civilization. There is a justifiable need to further evaluate this risk and evolve our understanding of the threat, mainly because the Yellowstone supervolcano has erupted three times over the past 2.1 million years. Another Yellowstone eruption would be like nothing humanity has ever experienced.

Existential Risks

Supervolcanoes like Yellowstone represent what is known as existential risks — ultra-catastrophes that could lead to global devastation, even human extinction. However, they are a rare but devastating event.

It is crucial to understand that what has happened before can happen again, eventually. Although we remain confined to our own lifetime's brief human time horizons, we treat improbable events as unreal and unimportant. In doing so, we leave ourselves vulnerable. We need to reimagine how to protect the future of humanity at risk.

Risk Group Efforts

If a supervolcano erupts, it could have a devastating blow to human civilization. Risk Group is evaluating whether applying emerging technologies to gather proactive, timely intelligence on when such an event would occur will impact how to mitigate the existential risks.

9. Cosmic Threats

How should we manage the cosmic threats that could impact the future of humanity?

Earth's Sustainability

There is a growing concern that cosmic events such as a nearby supernova, gamma-ray burst, near encounter with a large roaming planet, star, black hole, and more could directly wipe out all life on Earth. It could also disrupt the solar system so severely that the Earth would no longer be able to sustain life. [36]

It would be perhaps humanly impossible to prevent such an event. While experts believe we may be able to predict well-studied stars' behavior, there is a likelihood of either an unobserved and unidentified pair of white dwarfs/neutron stars or an unexplored binary system close enough crashing and going supernova results in devastating impact. The current technology would be helpless even with advanced knowledge to identify such unknowns.

Solutions to such cosmic problems would involve leaving the Earth. It also may be necessary to move a significant distance away from the entire solar system. For that to be a reality, we need to develop and far advance interstellar travel.

Potential Cosmic Events

While space is filled with wonders, it also brings enormous threats that could wipe out humanity. Since numerous possible space-based disasters could destroy humanity, it is crucial to evaluate all possible scenarios of cosmic events:

- Predatory Aliens: As plans are being made to advance our communication tools to reach other intelligent species in surrounding solar systems, there is a fear that once we make contact, they may try to destroy us. [37]

---

[36] Six cosmic catastrophes that could wipe out life on earth-The Conversation. https://theconversation.com/six-cosmic-catastrophes-that-could-wipe-out-life-on-earth-71178

- Asteroid Collision: The giant space rocks, commonly understood as asteroids when they hit the Earth, can bring enormous consequences to our species. Asteroids have hit before and can strike again. When the threat of an asteroid is severe, it is crucial to evaluate what can be done to protect our species from any encounter with an asteroid on a path of impact? Large can engender tsunamis, wildfires, and natural disasters and make the Earth uninhabitable. The question is, how likely is it to occur again, and do we have adequate technology and tools to manage the threat effectively?

- Solar Flares: are understood to be bursts of energy coming from a star. The Sun, a cauldron of condensed energy, can also emit tremendous radiation bursts in the form of solar winds. The question is, do we have a way of predicting g when or if these massive flares will happen again? And if it does, how can we protect the technological infrastructure in space and on Earth, maintaining our way of life—and perhaps existence? It is mostly when such a flare would eat away at the ozone layer, exposing us to deadly radiation, causing the degradation of ecosystems, including plankton, which supplies the majority of Earth's oxygen.[38]

- Gamma-Ray Bursts: are thought to occur during massive stars' deaths as they explode with incredible energy. While these bursts occur far enough away that their energies do not endanger Earth and its inhabitants, it is crucial to study that if a nearby burst were to occur, how we would protect the Earth and all its inhabitants.[39]

- Black Holes: The intense gravitational field surrounding black holes can devour all matters in their vicinity—perhaps even planets.[40]

---

[37] Is Stephen Hawking Right About Hostile Aliens? LIVESCIENCE. https://www.livescience.com/52439-stephen-hawking-hostile-aliens.html

[38] Sunspots and Solar Flares- NASA. https://spaceplace.nasa.gov/solar-activity/en/

[39] Gamma Ray Bursts-NASA. https://imagine.gsfc.nasa.gov/science/objects/bursts1.html

- Interstellar Dust Clouds: The possibility of our solar system hurtling through space and passing through dust clouds created by the remnants of the long-expired stars could plunge the Earth into another Ice Age. [41]

Our planet seems to have existed for 4.5 billion years. We need to understand the science behind each of these risks. We should manage and invest in preventing such cosmic events that could bring existential threats to humanity.[42]

Risk Group Efforts

Cosmic threats could have a devastating blow to human civilization. The Risk Group will continue to evaluate the risks to our species and planet. We will also assess how to apply emerging technologies to mitigate existential risks.

---

[40] What is a Black Hole? NASA.
https://www.nasa.gov/audience/forstudents/k-4/stories/nasa-knows/what-is-a-black-hole-k4.html

[41] Interstellar Dust Particles-NASA.
https://curator.jsc.nasa.gov/stardust/interstellardust.cfm

[42] These threats from outer space could end life on earth-Newsweek.
https://www.newsweek.com/earth-space-cosmos-threat-end-humanity-545075

10. Unknown-Unknowns

Are we taking the necessary steps to identify the most dangerous existential risks that are still not known?

Existential Risks

An existential risk is a risk that has the potential to eliminate all of humanity. It can, at the very least, impact and devastate large segments of the human population, leaving the survivors without sufficient means to rebuild society to current living standards. It is crucial to understand that most existential risks were natural until relatively recently. While responsible for significant progress and achievements, the last century's technological advances have taken us on a path of human-made existential risks. These new categories of technology triggered existential threats deserve much more attention.

Technology Triggered Threats

Humanity has outlasted threats for hundreds of thousands of years. While human civilization has faced a relatively low probability of existential catastrophic risks so far, the rapid technological growth has introduced a new class of existential threats to our collective future.

The gravest existential risks appear to be the ones that we only became aware of recently. As technology advances further, these new existential threats are likely to occur. When we generalize the technology trend, there might exist even more significant and consequential risks that we have not discovered yet. As a result, it is essential to put in place a risk identification mechanism, a system that helps identify unknown-unknowns of the existential risks in a timely manner.

The emerging technology-driven existential threats are already conceived; the only question is when and how the possibility of existential events triggered by self-replicating nanotechnology and super intelligent AI and so on will occur.

We must do a structured risk assessment of the emerging technologies' strategic security risks facing humanity. Although we have done a qualitative evaluation of potential and, in some instances, probable catastrophes, there is a need to put more resources into a quantitative assessment of existential risks and apply collective intelligence to build timely solutions.

Unknown Risk Implications

There is a need to make a distinction between the types of unknown-unknown risks. Natural versus human-made risks that are possible now but not identified versus the future unknown risks that are not identified. The existence of the natural and human-made, existing, and emerging unknowns of the risks needs to be our focus. To effectively identify any such risks requires an effective risk management framework and system that enables the collective identification process.

Human-made existential risks like nuclear war, artificial superintelligence, or self-replicating nanomachines are well within our ability to prevent. While our species faces greater existential peril than we ever have before, we now can protect ourselves by coming together and creating systems that can give us timely risk intelligence. However, more work is required to evaluate an existential catastrophe's probability more efficiently and accurately due to unknown risks. And for that, global coordination and cooperation would certainly help aid risk assessment and mitigation.

Risk Group Efforts

Since there is a need to apply technology to identify the risks in a timely manner, Risk Group proposes building an integrated security risk management system that will enable collective intelligence effort. Moreover, the use of smart sensors in all human domains will give us much needed intelligence on the unknowns of the technology risks. Much more needs to be done. But there must be a desire to work together to protect our collective future.

Risk Group is concerned about unknown future risks. With this Risk Report, we take the first step in acknowledging these risks. We will continue to bring future initiatives to strengthen our effort in protecting the future of humanity.

Conclusion

While we have briefly discussed the top ten strategic security risks facing humanity, many additional strategic security risks also require serious attention and preventive investment. Risk Group will continue its work to identify, evaluate, rate, and report all strategic security risks facing humanity. Risk Group will make a concerted effort to discuss each of these risks in the coming years.

An effective global governance system needs to be defined and designed to be enforceable, transparent, accountable, valid, legally binding, collaborative, and trust-driven to manage any strategic security risk.

While we have survived looking at the short-term risks over the years, we are now living in an age where it is critical to have a long-term view of the emerging strategic security risks that could negatively impact humanity's future. We must understand the strategic security risks ourselves, share insights with one another, and embed this strategic security risk thinking framework in our future generations if we are to survive as a human species.

There is no doubt that the coming years are going to be highly turbulent. We will need to learn to adapt and manage complex changes and solve formidable security problems. Moreover, we need to be resilient and develop resilient systems to face the emerging systemic shocks.

But more important than anything is a need to develop civilization-based thinking. It needs to be understood that tribalism is a destabilizing force. It discourages thinking and individual decision-making and forces loyalty to regressive causes and not forward-looking efforts. While tribal cohesion was essential to human survival for tribes at the beginning of civilization, today, as we develop artificial intelligence and begin to dream of exploring the universe, the question is whether we should think and act as a single human species or in silos as different tribes. How we represent our species and what comes next in our collective journey will entirely depend on whether we act as a cohesive civilization. I hope humanity can count on us.

About Risk Group's Founder And CEO

[Jayshree Pandya](#) (née Bhatt), Ph.D., a leading expert at the intersection of science, technology and security is the Founder and Chief Executive Officer of [Risk Group LLC](#), host of influential Risk Roundup Podcast/Vodcast, an award-winning Scientist, and a Futurist passionate about protecting the Future of Humanity.

This passion and dedication to protect the collective future have taken Dr. Pandya on a journey to understand the universe's language. Her quest for a security algorithm drives her involvement in a wide range of research to understand the core mechanisms by which the cosmos operates and where the existential risks emerge for the human species.

Her research across many domains has contributed to more than 100 publications, garnered her many advisory position honors, and is pursued to provide security solutions for humanity's future. Jayshree's inaugural book, [The Global Age: NGIOA @ Risk](#), was published by Springer in 2012. Her latest book, [Geopolitics of Cybersecurity](#), offers much-needed solutions to the rising digital disorder. Her upcoming books are on Cyber Warfare, Pandemics, and Artificial Intelligence.

She was [featured in CB Insights](#) in 2019.

Risk Group Call for Action

From hunter-gatherers to the first agrarian societies to the rise of vibrant and innovative civilizations around the world, our history demonstrates not only an ability but a skill at working together towards a better future. We carry the genes of our species' forward-thinking members: the ones who wisely chose to pool resources, work together, and overcome hardship rather than suffer in isolation and perish. These visionaries are our ancestors; we would not exist were it not for the wisdom of their choices.

Today, as we contend with a myriad of economic, political, technological, sociological, and planetary risks, Risk Group strongly believes that we must revisit our origins and collaborate at the individual and institutional levels. Like the existential threats our ancestors overcame thousands of years ago, our species is now at a crossroads where we can collectively ascend to the next chapter or lose everything that our forefathers fought so hard to build.

With this Risk Report, Risk Group is hereby calling individuals and entities across NGIOA to come together and collaborate to help research, review, rate and report the risks emerging from across nations in QCAGS. To join Risk Group efforts, please email Risk Group at info@riskgroupllc.com.

LIKE WHAT WE ARE DOING?

WE WOULD LOVE TO HEAR FROM YOU

Risk Group LLC

info@riskgroupllc.com   www.riskgroupllc.com

www.ingramcontent.com/pod-product-compliance
Lightning Source LLC
Chambersburg PA
CBHW061314140726
47998CB00006B/2394